Understanding Plant and Animal Cells

Understanding Plant and Animal Cells

Lillian Burton

ISBN-**10:** **09981454-0-8**
ISBN-**13:** **978-0-9981454-0-2**
Library of Congress Control Number: 2016916163
Lillian's Publishing, Olive Branch, MS

Table of Contents

Animal

Table of Contents (Continued)

Plant

Introduction

As a retired educator with 38 years of experience in public education, I am aware of how complicated science can be for some people. I have enjoyed much success using this while teaching my students, helping them achieve higher End of Course, college testing and other achievement test scores, and wanted to share it with everyone.

Therefore, I decided to produce a book that, hopefully, will continue to help students, their families, and educators expand knowledge about likenesses and differences among plant and animal cells.

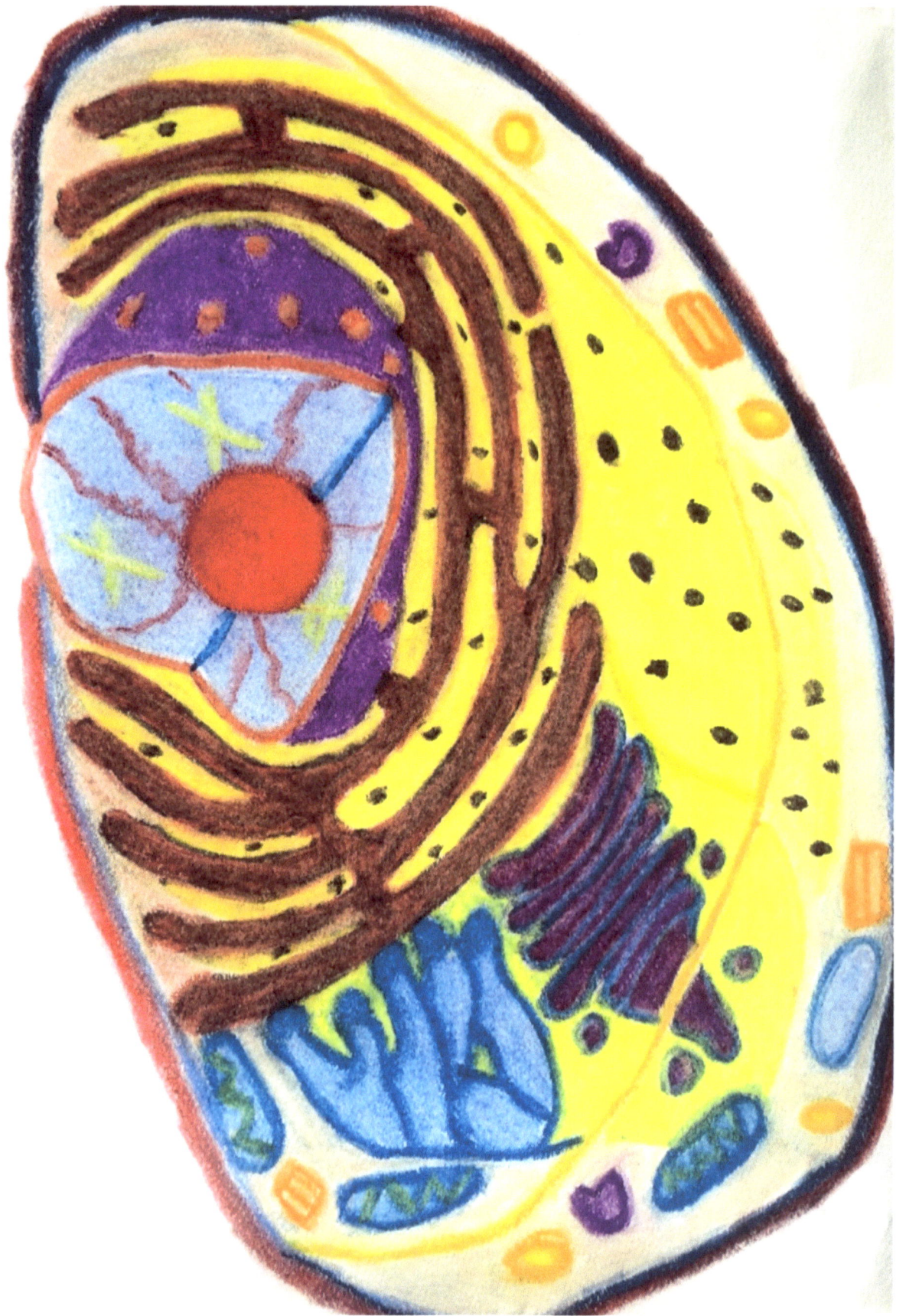

ANIMAL CELL

The animal cell described in this section are of organisms that take in oxygen and give carbon dioxide and water back into the atmosphere. They must consume their food so that cellular parts can perform certain functions. Those parts of the cell are called organelles. They are eukaryotic cells because they have more parts than the simpler cells called prokaryotic.

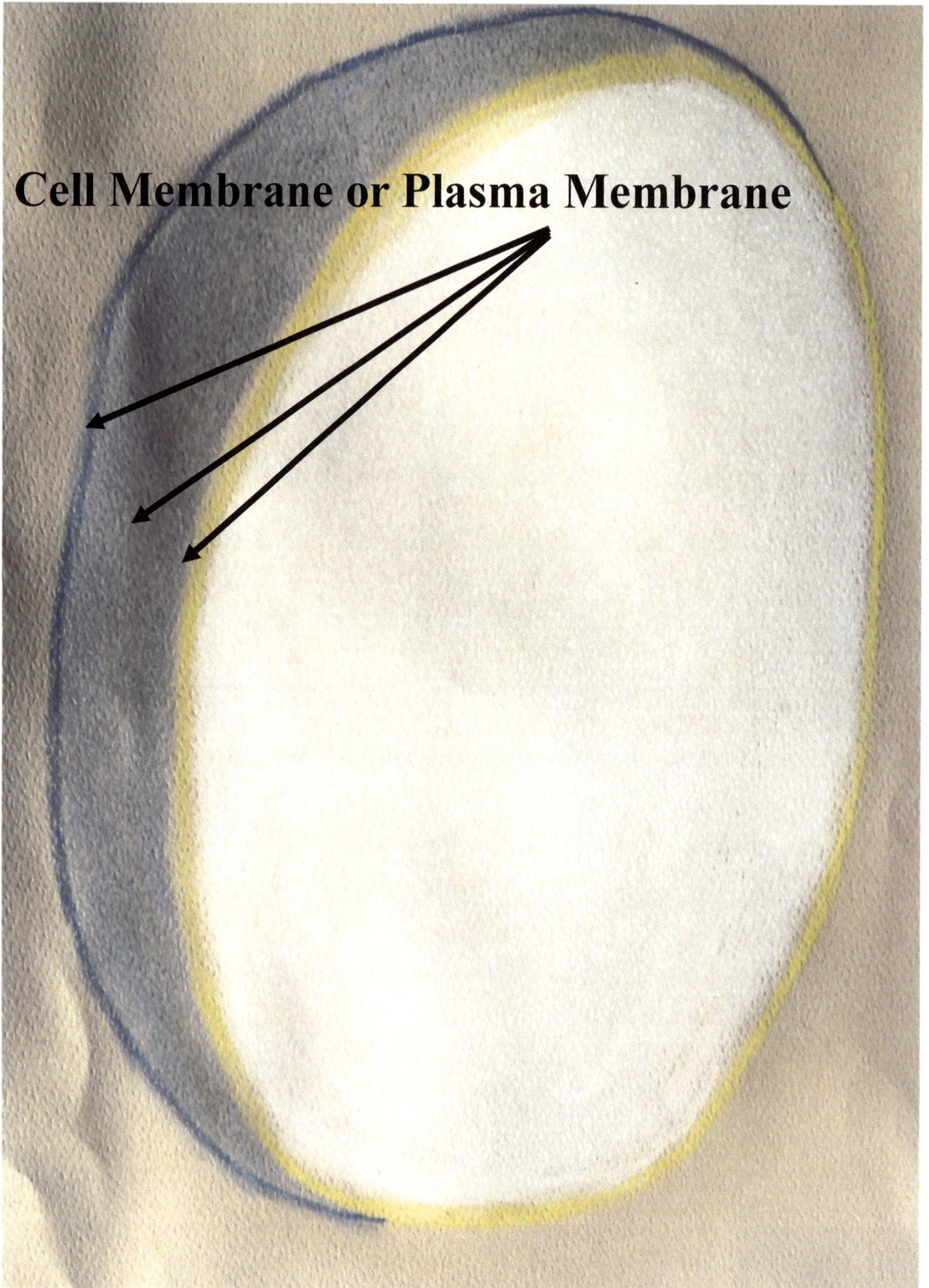

Cell Membrane or Plasma Membrane

Cell Membrane or Plasma Membrane – surround the cell and controls what nutrients passes into and waste that leaves the cell.

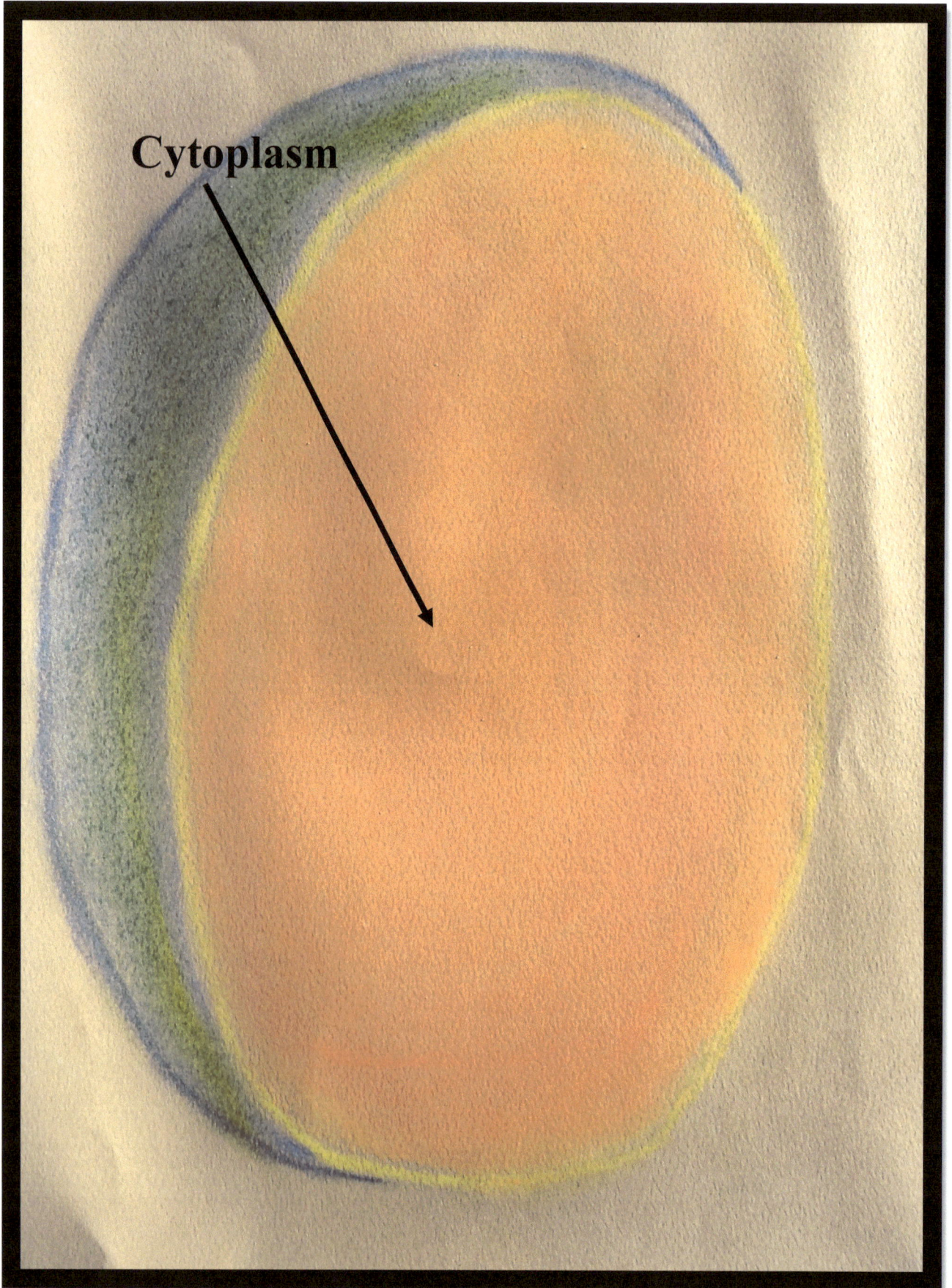

Cytoplasm

Cytoplasm – fluid-like living material within a cell, which holds many dissolved chemicals.

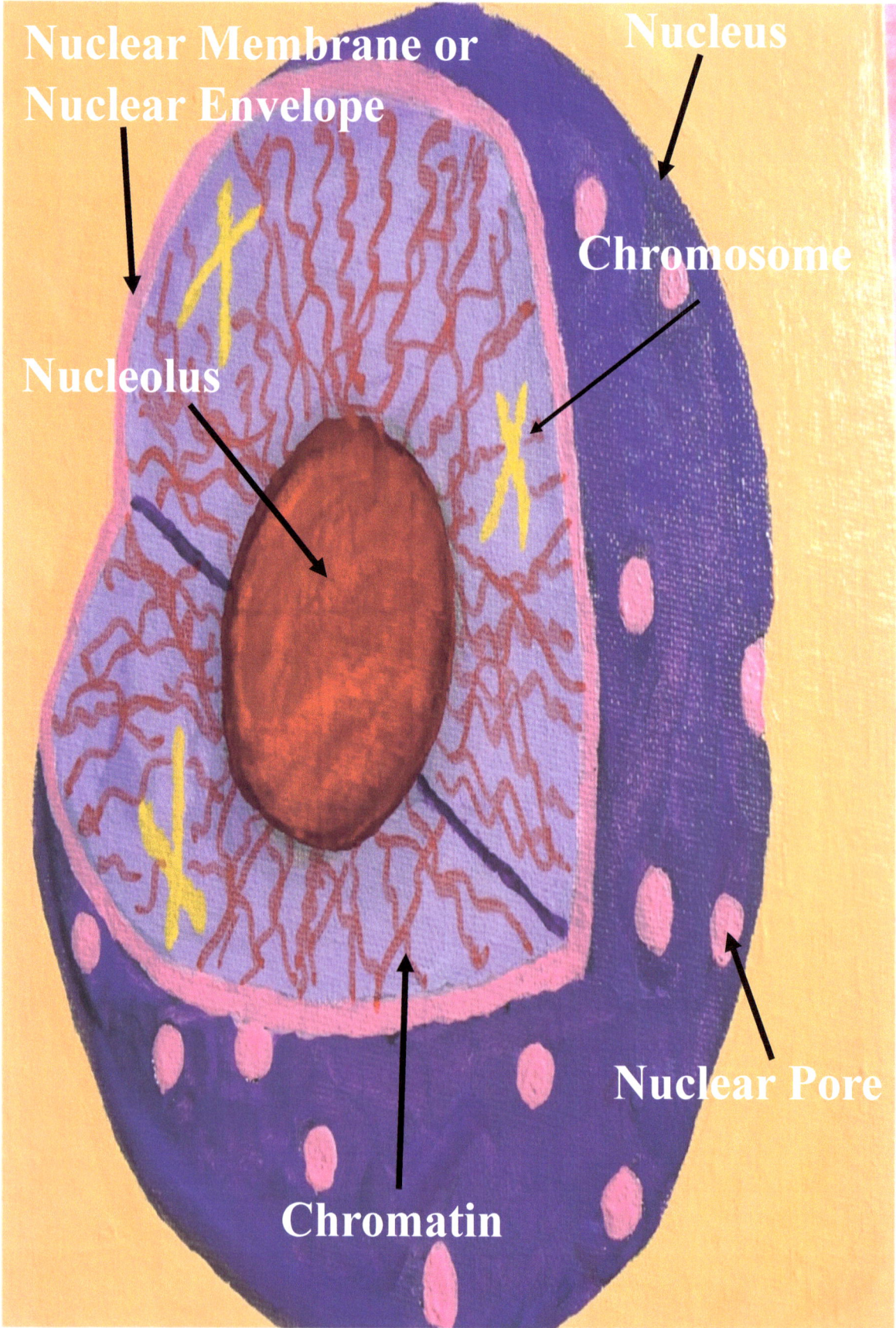

Nuclear Membrane or Nuclear Envelope

Nucleus

Chromosome

Nucleolus

Chromatin

Nuclear Pore

Nucleus – contains almost all of the cell's DNA for making protein and genetic material. It controls what occurs in the cell.

Nuclear Membrane or Nuclear Envelope – has two membranes that surround the nucleus. It has thousands of pores through which proteins and other molecules move into and out of the nucleus to other parts of the cell.

Nucleolus – is a small dense area in the nucleus. It contains DNA and RNA, and it makes ribosomes.

Nuclear Pore – an opening that allows for the entry and exit of substances into and out of the cell.

Chromatin – is located throughout the nucleus, and consists of DNA. DNA bonds to proteins which are sorted into packages to control cell division.

Chromosomes – are thread-like structures that contain genetic (DNA) information which is passed from one generation to the next.

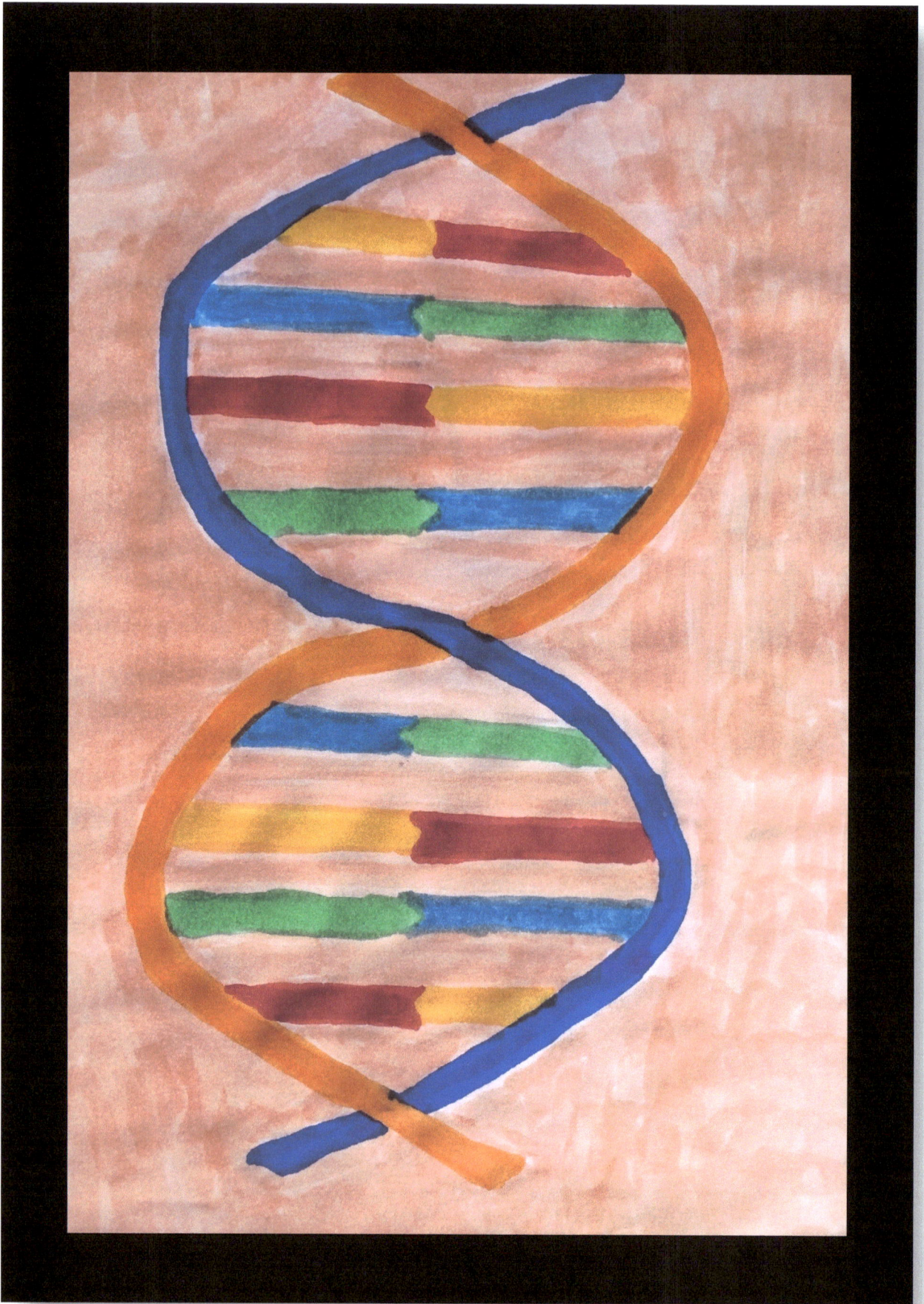

DNA – has coded instruction for making protein and other important molecules.

Ribosomes – are on the surface of Rough ER and makes protein. Ribosomes are mini particles of RNA and proteins found in the cytoplasm. They use a code from the nucleus to make proteins.

Rough Endoplasmic Reticulum (Rough ER) – an internal long sac-like membrane that lies close to the nucleus. Rough ER makes the lipid parts of the cell membrane. It has ribosomes on its surface that are involved in synthesizing proteins. New proteins leave the ribosomes, and the Rough ER chemically changes them.

Smooth Endoplasmic Reticulum (Smooth ER) – accepts ribosomes from the Rough ER and distributes them to their special locations inside the cell and also transports proteins.

Golgi Apparatus – arranges protein molecules and transports them to different locations within the cell.

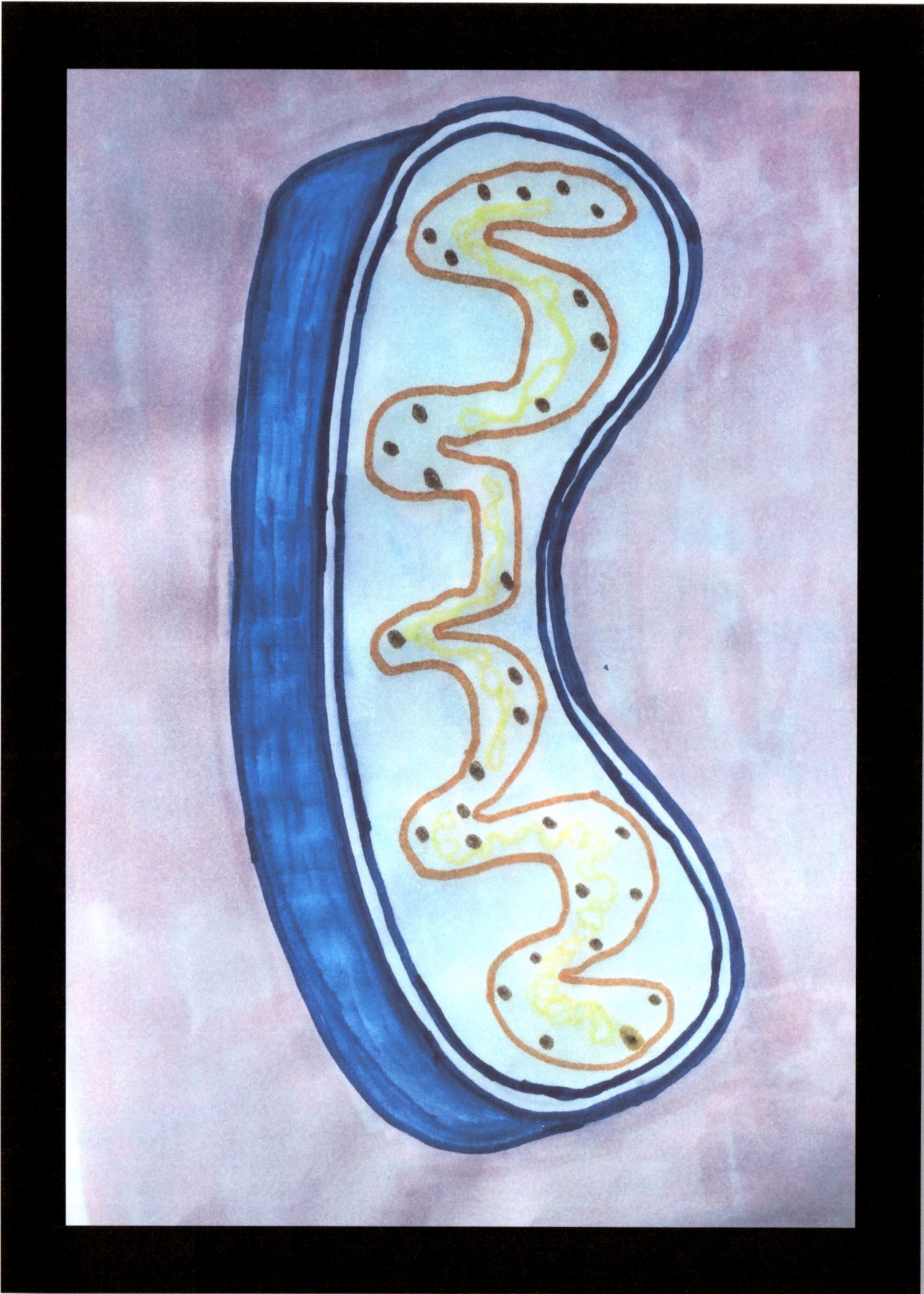

Mitochondria – produces chemical energy for the cell, the powerhouse, from food and where cell respiration occurs.

Lysosome – consist of powerful enzymes that digest trapped food particles and protects the body from harmful invaders.

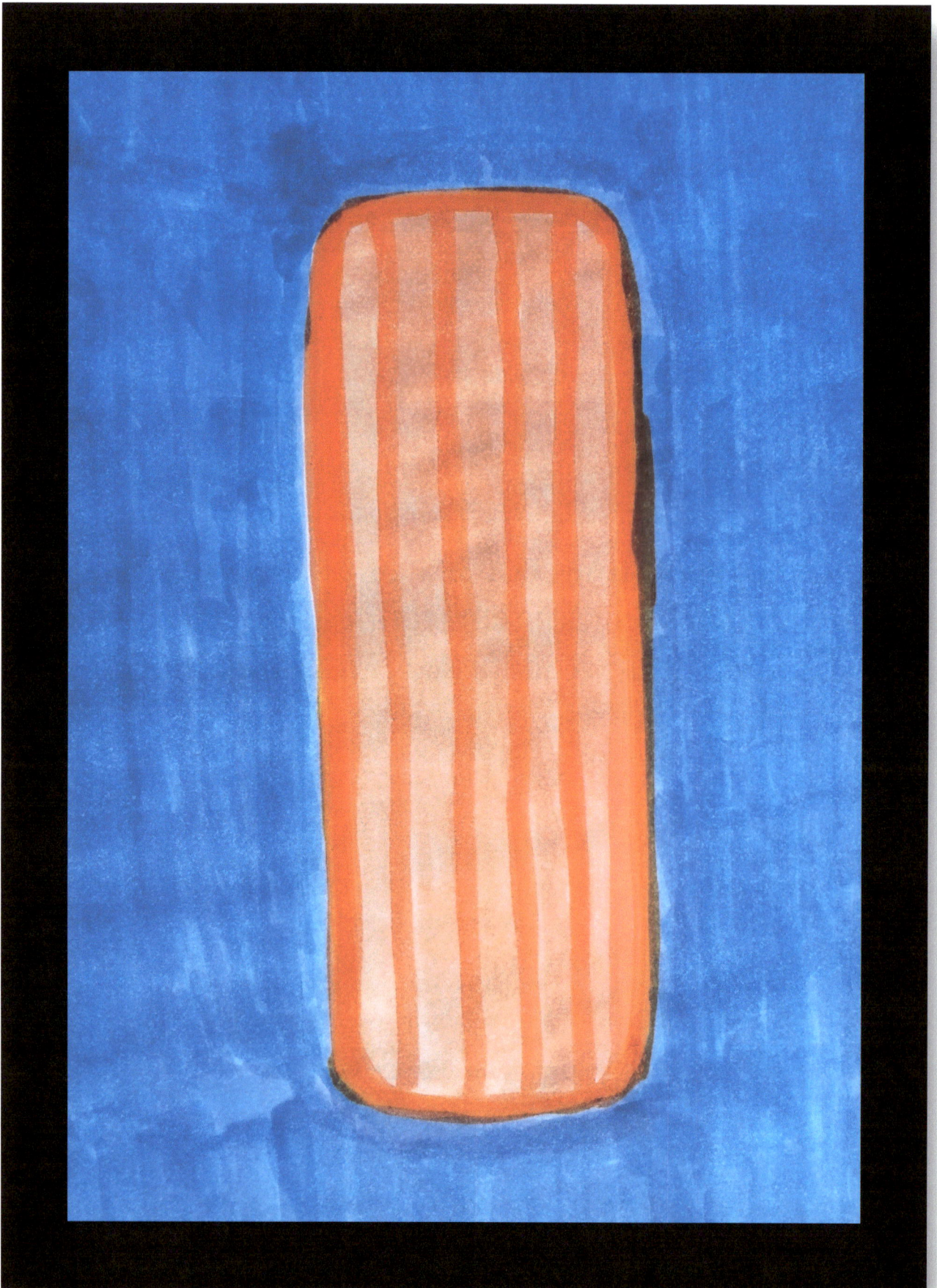

Centriole – helps distribute chromosomes to daughter cell during cell division.

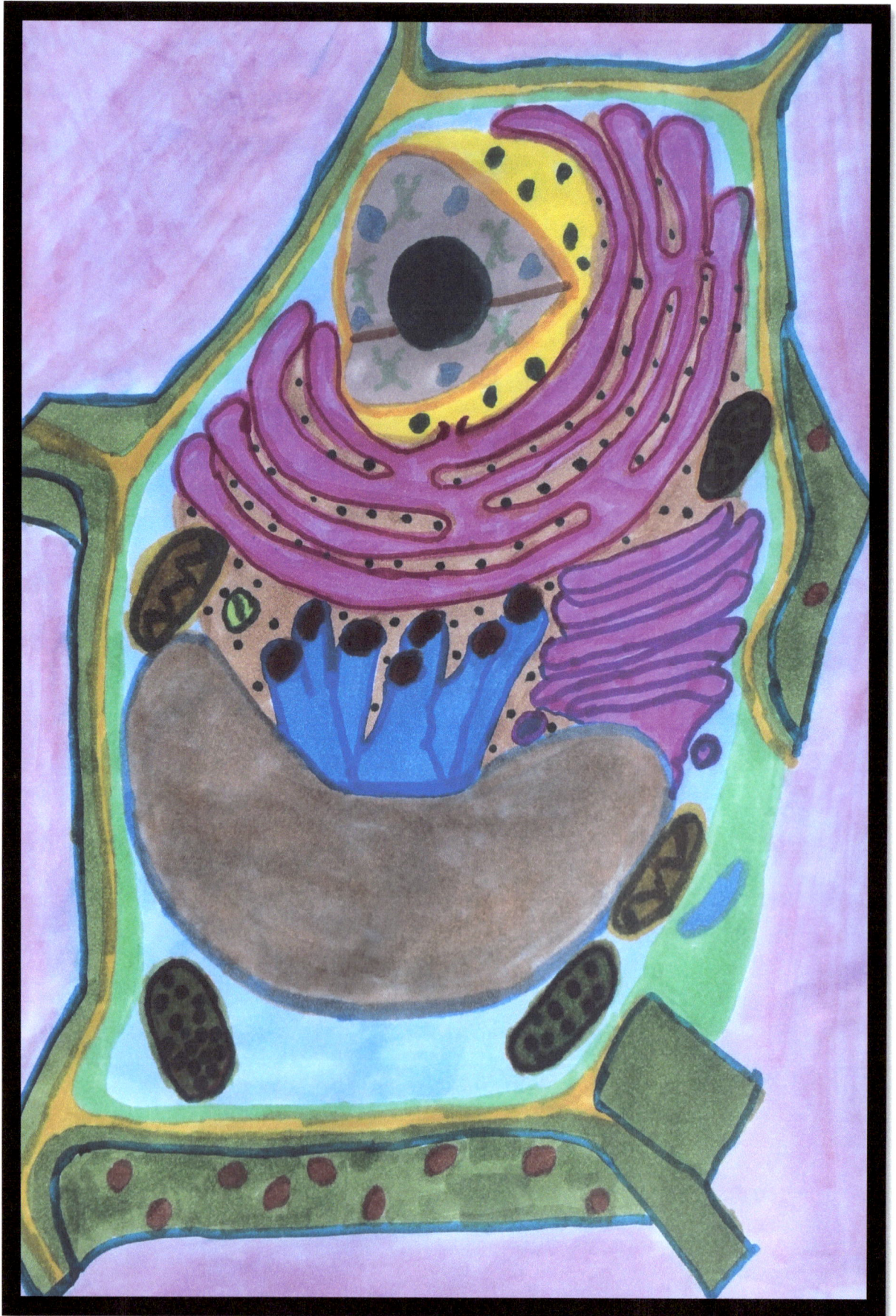

PLANT CELLS

Plant cells described in this section are of organisms that take in water, carbon dioxide, and sunlight from the atmosphere. These substances are used to produce their own food which makes them producers. They give oxygen back to the atmosphere and they are food for animals. The cellular parts are also referred to as cell organelles just as they are in animal cells. They are called eukaryotic cells because they have more parts than other living things made of simpler cells called prokaryotic.

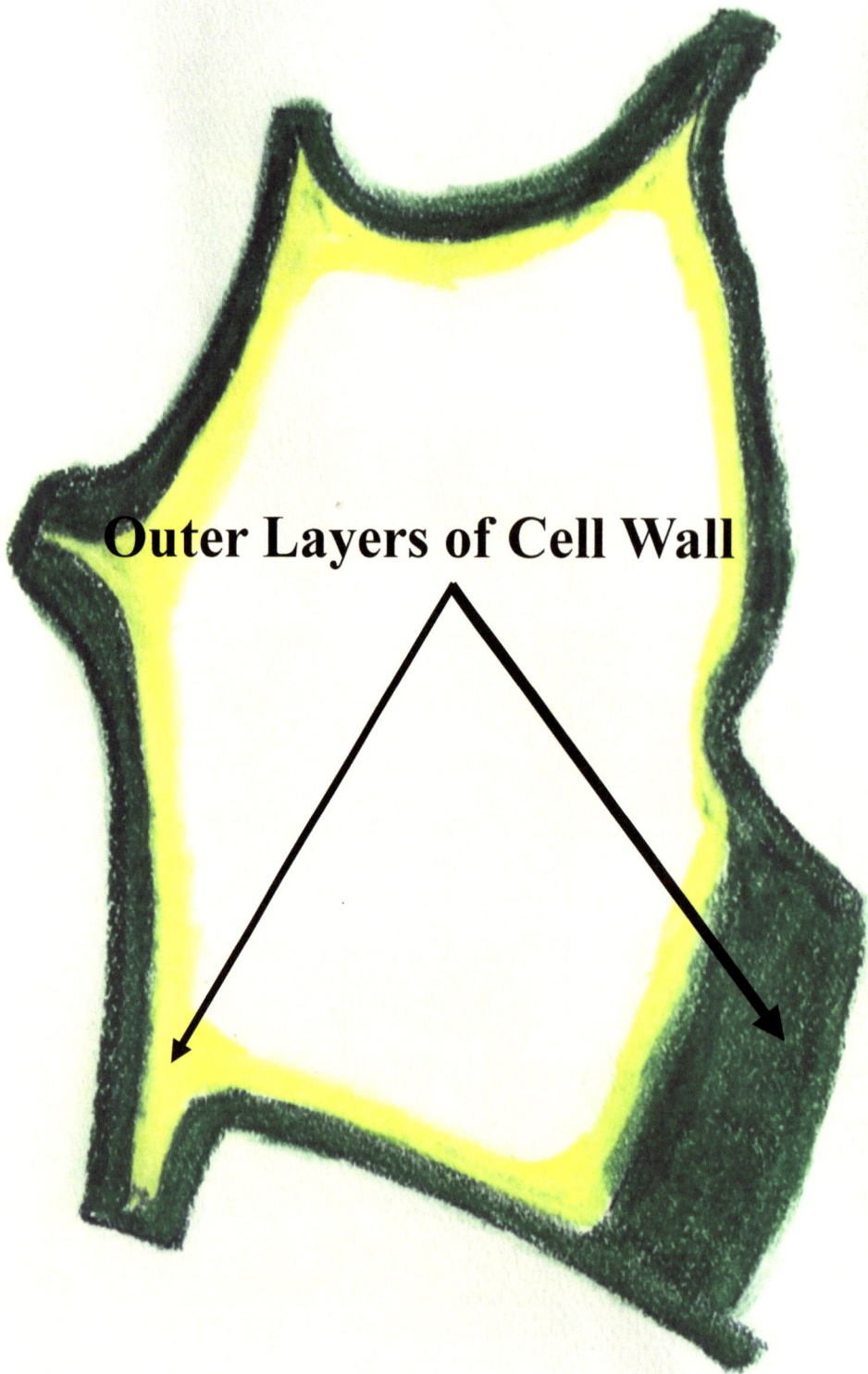

Outer Layers of Cell Wall

Cell Wall – consist of two tough outer layers located outside of the cell membrane that protects the cell and helps it exchange materials with the environment. It gives strength and support to the plant cell.

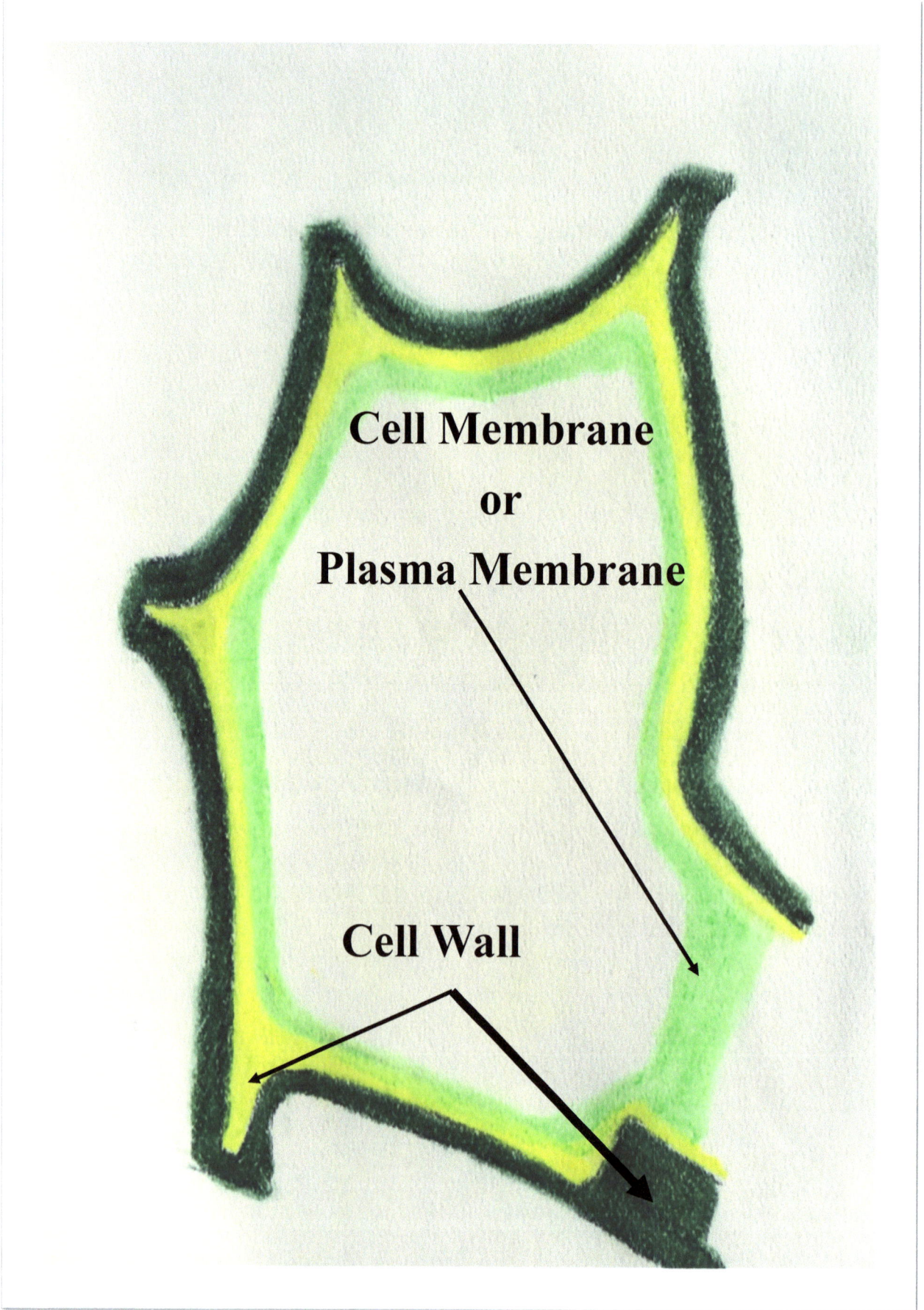

**Cell Membrane
or
Plasma Membrane**

Cell Wall

Cell Membrane or Plasma Membrane– is located inside of the cell wall. It is a barrier or fence that holds the material inside of the cell. It controls what passes into and exits the cell.

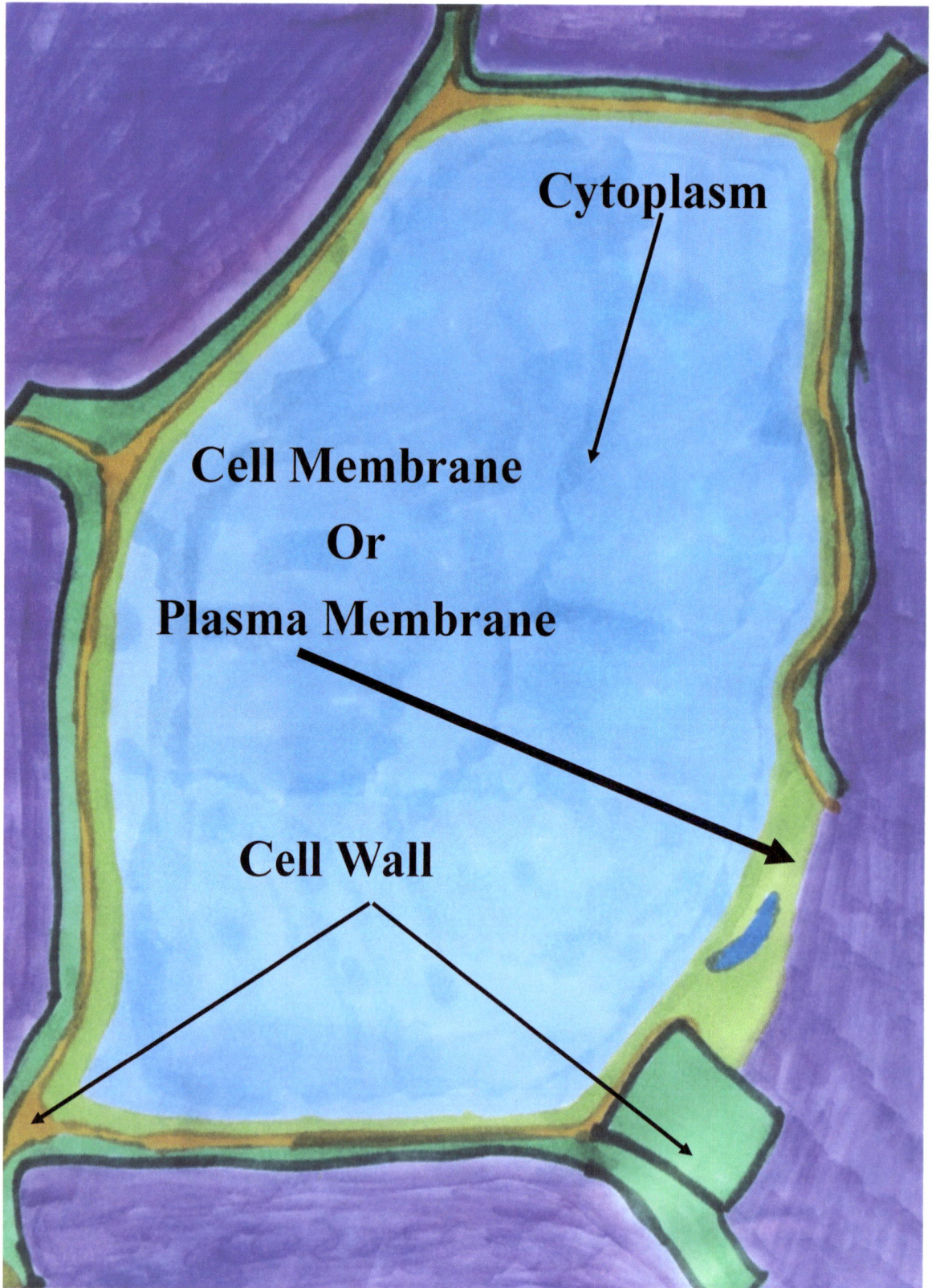

Cytoplasm

Cell Membrane
Or
Plasma Membrane

Cell Wall

Cytoplasm – fluid-like living material inside of a cell, which holds different dissolved substances.

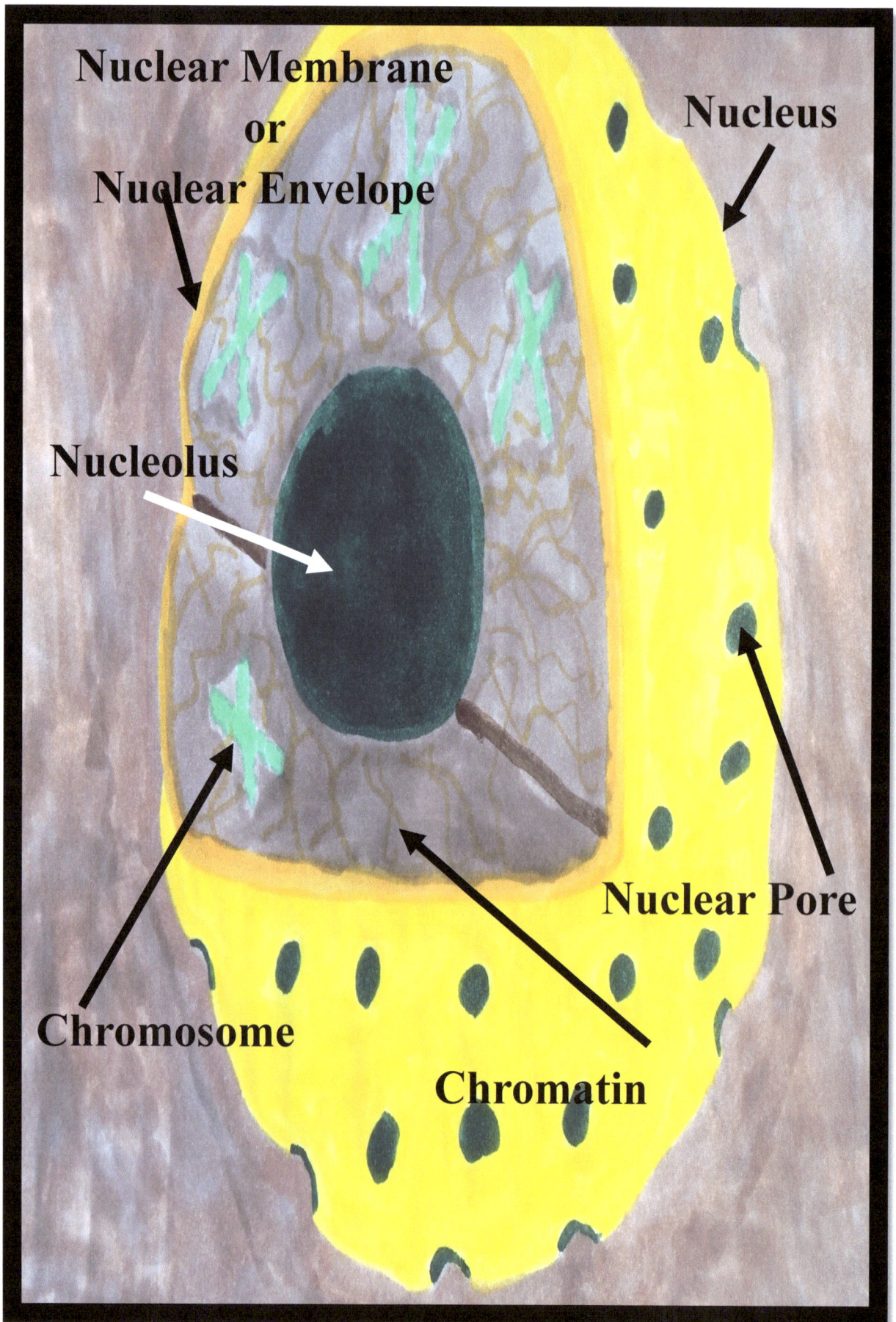

Nuclear Membrane
or
Nuclear Envelope

Nucleus

Nucleolus

Chromosome

Nuclear Pore

Chromatin

Nucleus – It is the control center of the cell, hereditary material and makes protein from DNA.

Nuclear Membrane or Nuclear Envelope – has two membranes that hold materials inside of the nucleus. It has pores through which proteins and other molecules move into and out of the nucleus to other parts of the cell.

Nucleolus – It makes ribosomes. It is composed of protein, DNA, and RNA. It changes DNA into three types of RNA which are transported outside of the nucleolus to be used to perform different tasks such as making protein and changing coded material.

Nuclear Pores – an opening that allows for the entry and exit of material in the cell.

Chromatin – consist of DNA that is bonded to protein and is throughout the nucleus. It controls cell division.

Chromosomes – thread-like structures that contain genetic (DNA) code which is passed on from one generation to the next.

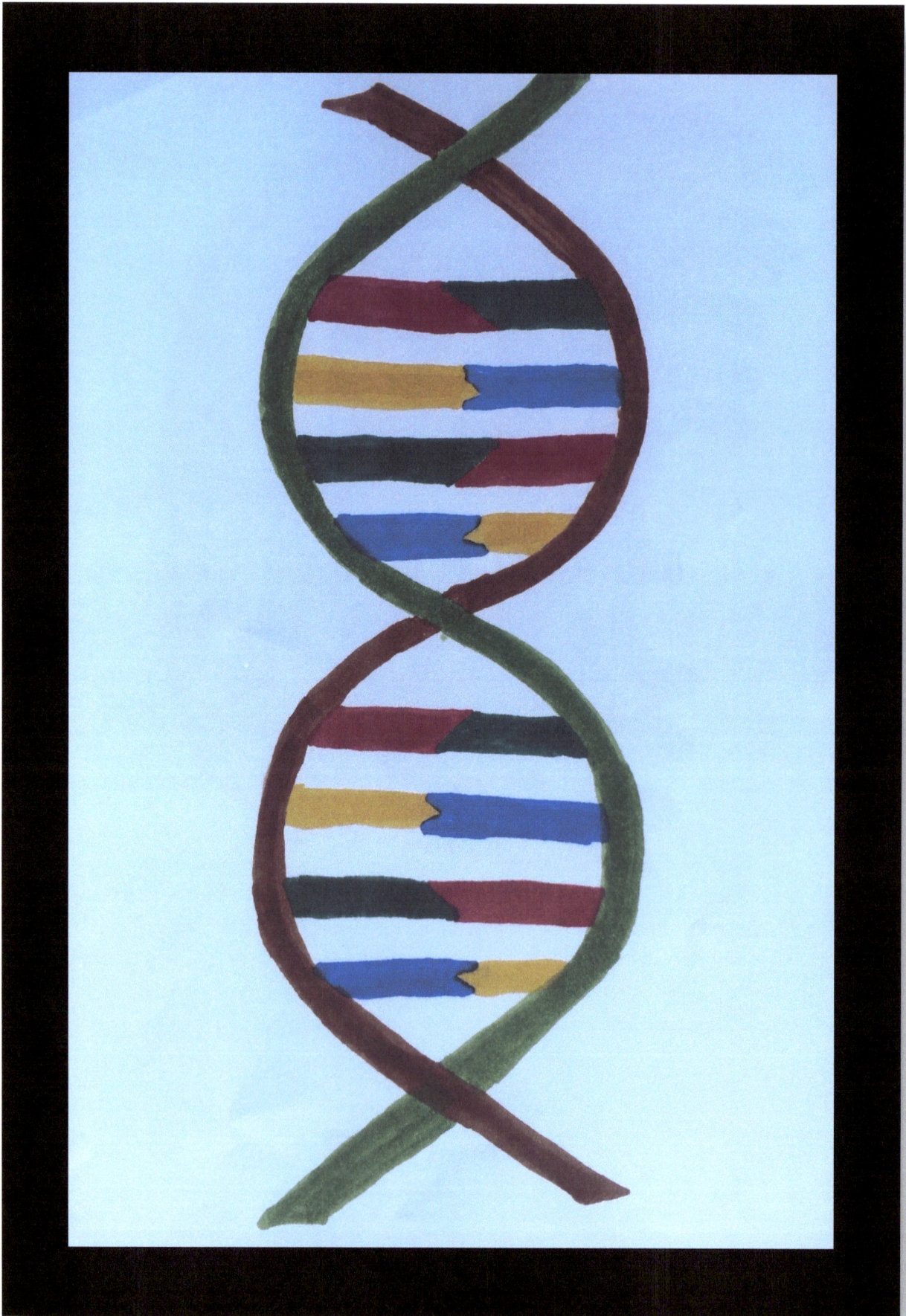

DNA – has coded instruction for making protein and other important molecules.

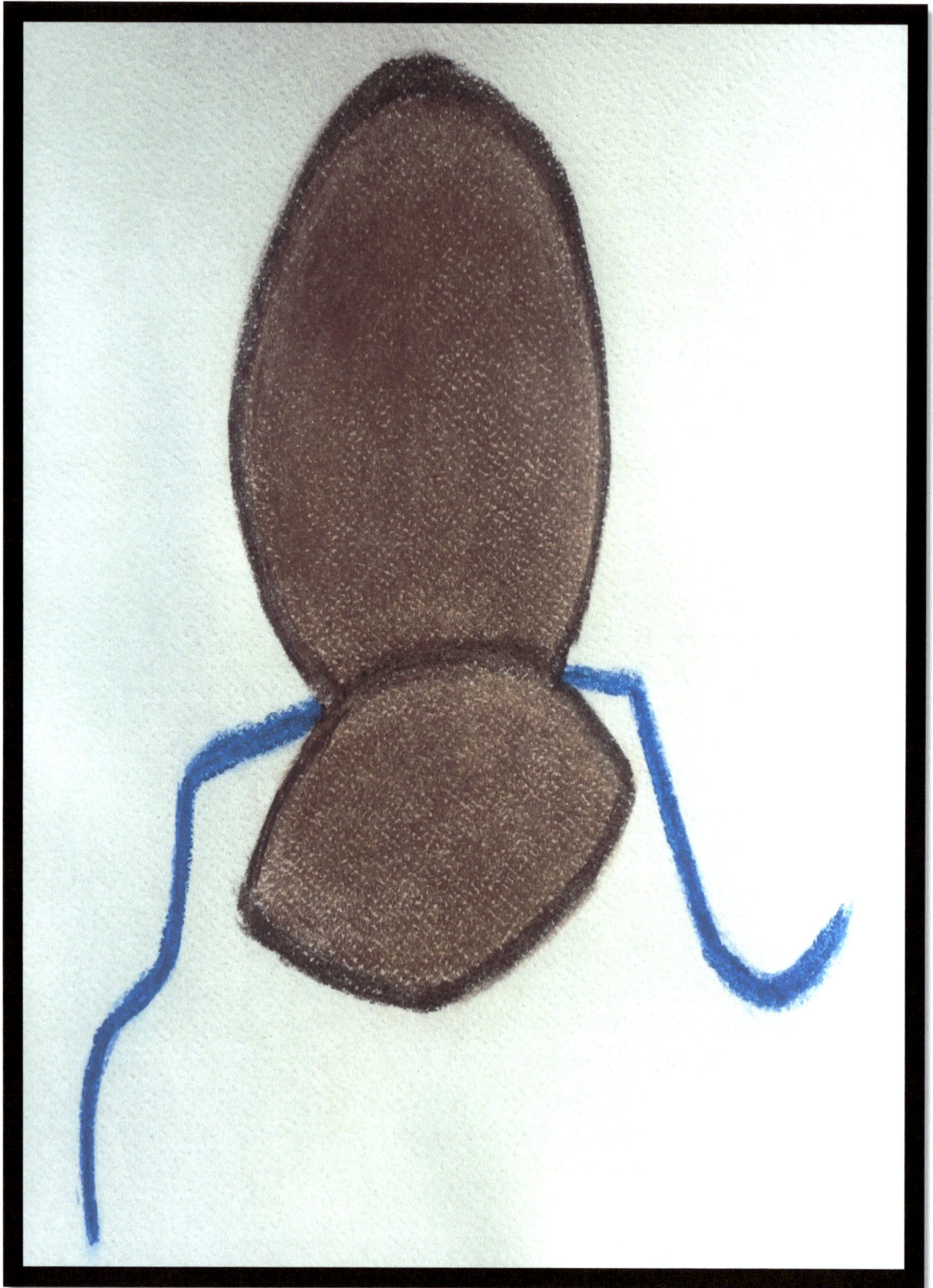

Ribosomes – are on the surface of Rough ER and makes amino acid that creates a certain kind of protein. Ribosomes are mini particles of RNA and proteins found all around the cytoplasm. They use a code from the nucleus to make proteins.

Rough Endoplasmic Reticulum (Rough ER) – a long interconnected sac that lies close to the nucleus and has ribosomes on its surface that are involved in synthesizing proteins.

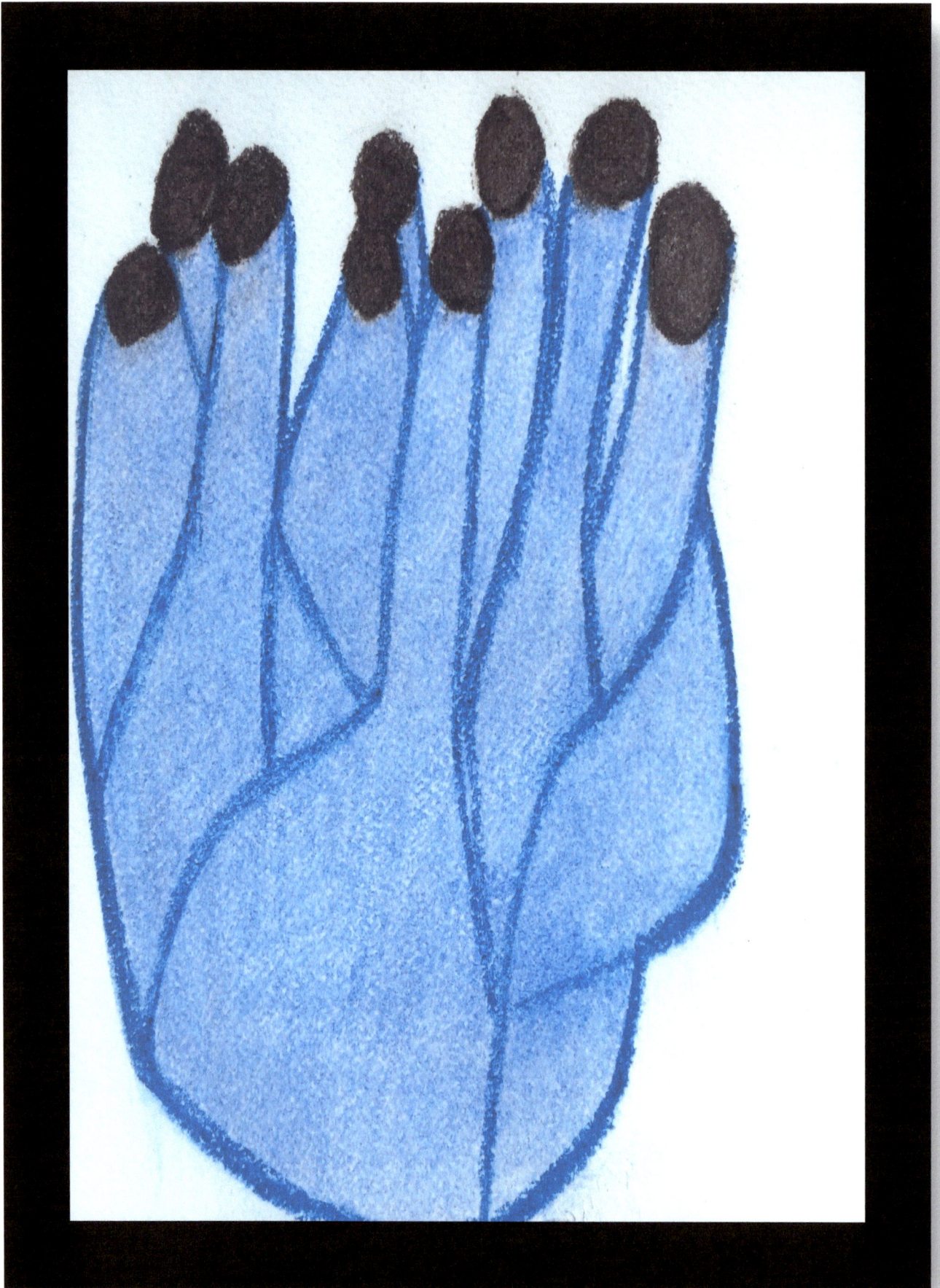

Smooth Endoplasmic Reticulum (Smooth ER) – accepts ribosomes from the Rough ER and moves protein and carbohydrates to the Golgi apparatus, cell membrane, and other areas where it is needed.

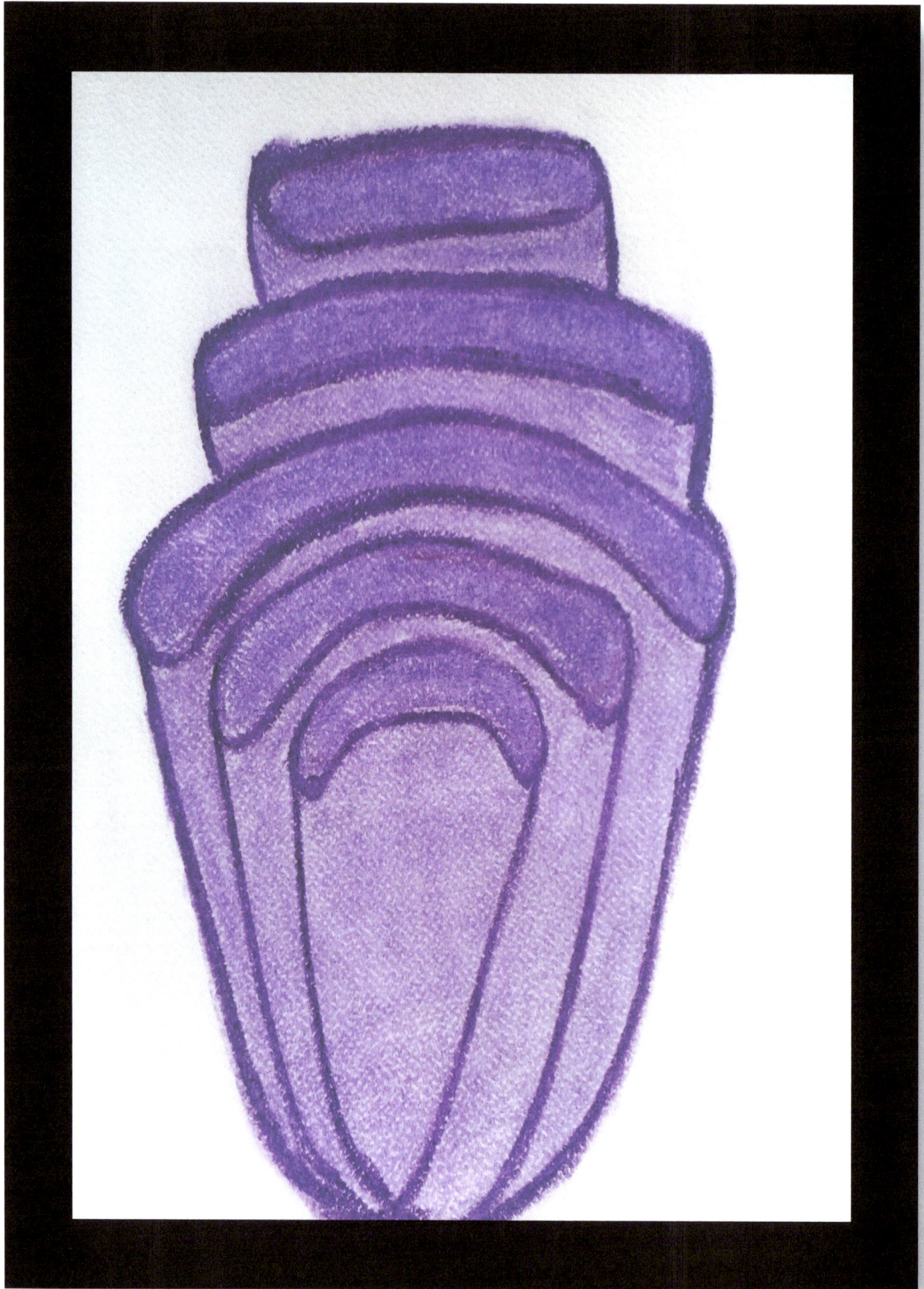

Golgi Apparatus – changes, sorts, and package materials together as needed by different functions in the cell.

Mitochondria – changes some of its oxygen and sugar into ATP which is chemical energy needed by the cell.

Vacuole – a large storage area, for food and other materials that recently entered the cell. It also maintains structure and shape of the cell.

Vacuoles exist in a variety of shapes.

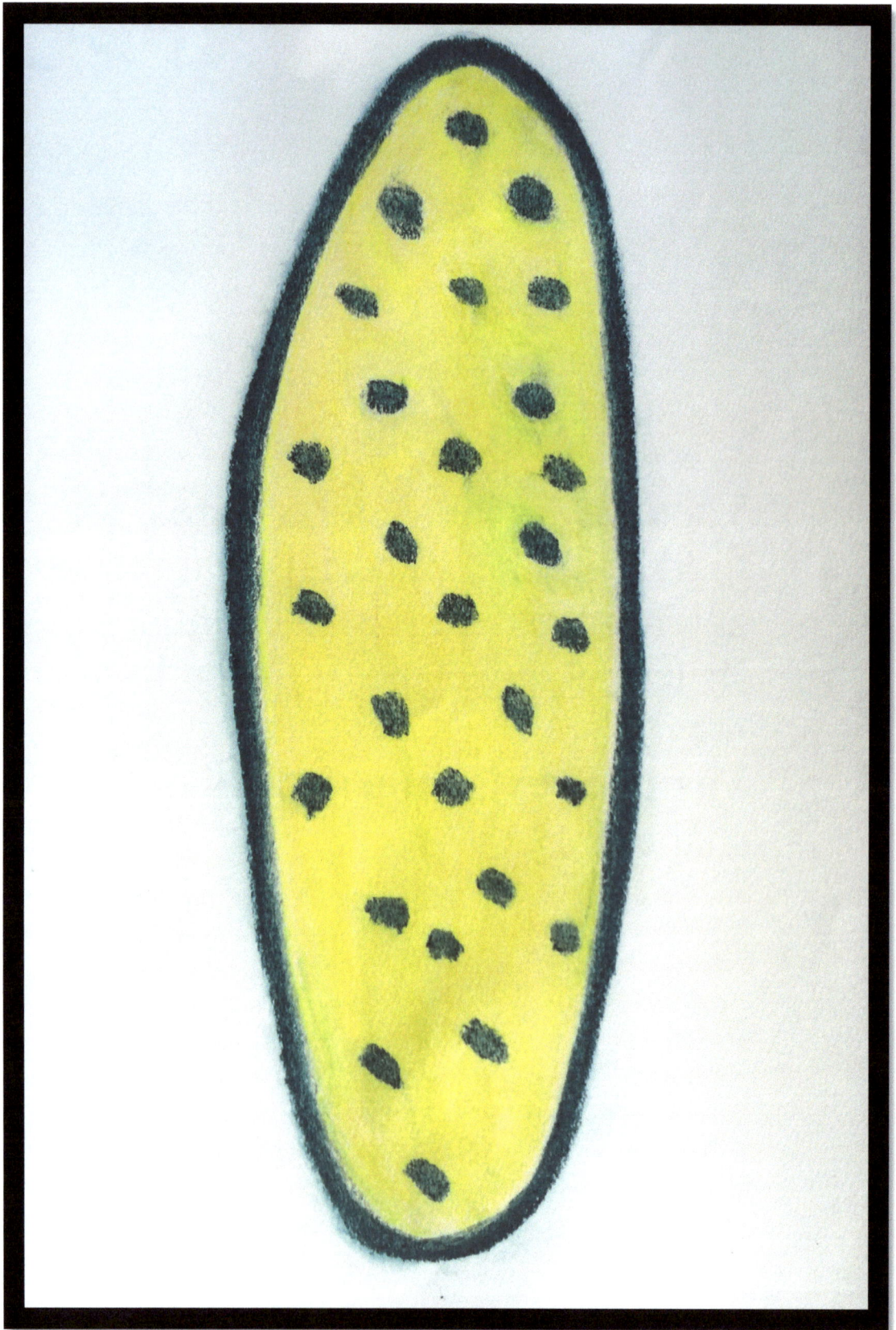

Chloroplast – where photosynthesis occurs. It helps plants to make their own food by changing the Sun's energy into sugars.

Index

Resources

Graham, Keith. 1986. *Biology, God's Living Creation*. Pensacola, Fla. (Box 18000, Pensacola 32523-9160): Beka Book.

Miller, Kenneth R and Joseph S Levine. 2010. *Miller & Levine Biology*. Boston, Mass.: Pearson.

http://www.plant-biology.com/

Thanks for purchasing this book. If there are any other life science/biology related topics that you like to see published in a book please send an email to:
lpublishing7781@gmail.com